Who's Making All That NOISE?!

Darrin Lunde
Illustrated by Erica J. Chen

Charlesbridge

It's a loud hammering on a tree.

A woodpecker.

Woodpeckers hammer trees
to find insects to eat.
They also hammer
to call other woodpeckers.

A dolphin.

Who's making all that noise?!

It's a rumbling like thunder.

A lion.

Who's making all that noise?!

It's coming from the edge of a pond.

A quacking frog.

Have you ever heard a frog quack like a duck? These frogs live in Australia, and the males quack to attract a mate.

An elephant.

An elephant will use its nose like a trumpet when it is scared or angry.
It trumpets to warn other elephants.

CH-CH-CH-CH-CH-CH-CH-CH-CH-CH-CH . . .

Something is rattling on the ground.

A rattlesnake.

Rattlesnakes shake a rattle on the tip of their tail. It is a warning—stay away!

A human.

People are some of the noisiest animals. What do you have to say about that?

SOUNDS FUN

Grunts are an ocean fish that make a loud groaning sound when distressed. This helps warn other grunts of danger.

Male gorillas beat their chest to make a drumming sound that attracts females and scares off rivals.

Sharks do not have any organs that can produce sound. This gives new meaning to the phrase "silent but deadly."

Male club-winged manakin birds of South America flap their wings to make a buzzing sound that attracts females.

Snapping shrimp use their claws to make a powerful snap that stuns small fish and worms. The shrimp then eats the stunned prey.

The male white bellbird of South America makes the loudest call of any bird. It is as loud as a jackhammer and can sound like a bell or a siren.

Male crickets chirp by rubbing the edges of their wings together. The females approach only the best-sounding males.

Male and female gibbons howl different parts of a complex song. To a human listener, a gibbon duet can sound both sad and beautiful.

For ALL of my noisemakers: Sakiko, Sakura, Asahi, and Midori. —D. L.

For Stella, Ginger, Yen, and Joanne. You're all number 1 to me. —E. J. C.

Text copyright © 2025 by Darrin Lunde
Illustrations copyright © 2025 by Erica J. Chen
All rights reserved, including the right of reproduction in whole or in part in any form. Charlesbridge and colophon are registered trademarks of Charlesbridge Publishing, Inc.

At publication, all URLs in this book were accurate. Charlesbridge, the author, and the illustrator are not responsible for the content of any website.

Charlesbridge • 9 Galen Street, Watertown, MA 02472
www.charlesbridge.com

Library of Congress Cataloging-in-Publication Data
Names: Lunde, Darrin P., author. | Chen, Erica, illustrator.
Title: Who's making all that noise?! / Darrin Lunde; illustrated by Erica Chen.
Description: Watertown, MA: Charlesbridge, [2025] | Audience: Ages 3-7 | Audience: Grades K-1 | Summary: "Guess the noisemaker from the noise and learn what sounds animals make—and why."—Provided by publisher.
Identifiers: LCCN 2024013066 (print) | LCCN 2024013067 (ebook) | ISBN 9781623546267 (hardcover) | ISBN 9781632892720 (ebook)
Subjects: LCSH: Animal sounds—Juvenile literature. | Animal communication—Juvenile literature. | Animal behavior—Juvenile literature.
Classification: LCC QL765 .L86 2025 (print) | LCC QL765 (ebook) | DDC 591.59/4—dc23/eng/20240813
LC record available at https://lccn.loc.gov/2024013066
LC ebook record available at https://lccn.loc.gov/2024013067

Printed in China • OPIC
(hc) 10 9 8 7 6 5 4 3 2 1

Illustrations done digitally
Text type set in P22 Stanyan Autumn
Edited by Alyssa Mito Pusey
Designed by Ellie Erhart
Production supervised by Mira Kennedy